ESSENTIALS OF HYDROGEN MEDICINE FOR ADULTS

A Guide to Integrative Hydrogen Therapy **revolution** for Vibrant Living, Beyond Conventional Medicine and Embracing a New Paradigm of Wellness and optimal health

Dr Sandra Almeida

Table of contents

INTRODUCTION

Background of Hydrogen Medicine

The utilization of hydrogen in medical applications, often referred to as hydrogen medicine, has gained significant attention in recent years. Originating from its fundamental role as an element, hydrogen exhibits unique properties that make it a potential therapeutic agent. With its smallest and simplest molecular structure, hydrogen can penetrate cellular barriers, presenting opportunities for novel medical interventions.

This section explores the historical context and scientific foundations that led to the exploration of hydrogen as a medicinal tool. It delves into early research, showcasing the gradual evolution of hydrogen medicine from theoretical concepts to practical applications in healthcare.

The exploration of hydrogen as a therapeutic agent has its roots in the understanding of its unique molecular properties. Hydrogen, the lightest and simplest element, consists of a single proton and electron. This simplicity, coupled with its small size, enables hydrogen molecules to easily penetrate cellular structures, potentially influencing various physiological processes. The concept of hydrogen as a medical intervention gained momentum as researchers began to recognize its potential as an antioxidant and anti-inflammatory agent.

Historically, hydrogen's medicinal properties were noted in the context of its application as an inhalation gas to treat decompression sickness in deep-sea divers. However, it wasn't until the early 21st century that scientific interest surged, driven by discoveries related to its impact on oxidative stress and cellular function. This section delves into pivotal moments in the development of hydrogen medicine, tracing its evolution from an intriguing concept to a subject of rigorous scientific investigation.

The primary purpose of this book is to serve as a comprehensive guide to hydrogen medicine for adults, offering a balanced blend of scientific insights and practical applications. In a rapidly evolving field, the book aims to distill complex research findings into accessible information, catering to both medical professionals and individuals keen on optimizing their health.

The scope encompasses various dimensions of hydrogen medicine, starting with an exploration of its molecular properties and extending to its diverse applications in healthcare. By elucidating the potential benefits across different health domains, the book strives to provide readers with a nuanced understanding of how hydrogen can be integrated into existing health regimens.

The book is not merely a compilation of scientific data but a practical resource that equips readers with the knowledge to make informed decisions. Whether readers seek to enhance their understanding of cutting-edge medical interventions or explore complementary approaches to health, the book aspires to be a comprehensive reference guide.

The purpose of this book is to provide a comprehensive understanding of hydrogen medicine for adults. It aims to bridge the gap between scientific research and practical knowledge, offering insights into the potential benefits and applications of hydrogen in healthcare. From its molecular properties to clinical studies, the book aspires to serve as a go-to resource for both medical professionals and individuals seeking to explore hydrogen as a complementary approach to well-being.

The scope encompasses various aspects, including the physiological effects of hydrogen, methods of administration, clinical studies, and practical integration into daily life. By providing a holistic view, the book intends to empower readers with the knowledge needed to make informed decisions about incorporating hydrogen medicine into their health regimens.

The diverse target audience reflects the inclusive nature of the book. Medical professionals, including doctors, researchers, and clinicians, form a primary readership. For these professionals, staying abreast of emerging therapeutic modalities is crucial, and this book aims to provide a consolidated resource for understanding hydrogen's role in medicine.

Simultaneously, the book caters to a broader audience, including health-conscious individuals, wellness enthusiasts, and those with a general interest in science and medicine. The content is structured to be informative without overwhelming, making it accessible to readers with varying levels of scientific expertise.

Ultimately, the target audience spans those seeking evidence-based insights into hydrogen medicine, recognizing its potential contributions to health and well-being. By addressing a diverse readership, the book endeavors to contribute to the democratization of knowledge in the field of hydrogen medicine.

This book is tailored for a diverse audience interested in the intersection of science, medicine, and well-being. The primary target audience includes healthcare professionals such as doctors, researchers, and clinicians looking to stay updated on emerging trends and potential therapeutic interventions. Additionally, the book caters to curious individuals seeking accessible information about hydrogen medicine, encouraging proactive engagement with their health.

The content is designed to be informative yet approachable, making it suitable for readers with varying levels of scientific background. Whether you are a seasoned healthcare professional or a health-conscious individual, the aim is to empower you with knowledge that facilitates a nuanced understanding of hydrogen's role in medicine.

By defining the background, purpose, and target audience clearly, the introduction sets the stage for a journey through the intricate world of hydrogen medicine, ensuring readers grasp the significance and relevance of the subsequent chapters.

CHAPTER ONE

Understanding Hydrogen

Hydrogen, the lightest and most abundant element in the universe, holds a pivotal role in both cosmology and chemistry. Its atomic structure consists of a single proton and a lone electron, making it the simplest and smallest element. In the periodic table, hydrogen takes the topmost position, symbolized by the letter 'H.' This fundamental element is colorless, odorless, and tasteless, existing in various forms such as molecular hydrogen (H_2) and atomic hydrogen (H).

In its natural state, hydrogen is a diatomic molecule, meaning it typically pairs with itself to form H_2.

This diatomic form is stable and represents the molecular hydrogen commonly associated with medicinal applications. Hydrogen's simplicity is a key factor in its potential as a therapeutic agent, as it allows for easy penetration of biological barriers, including cell membranes.

Molecular Hydrogen

Molecular hydrogen, denoted as H_2, is at the forefront of hydrogen medicine research. This diatomic molecule consists of two hydrogen atoms covalently bonded, forming a stable and neutral entity. Its significance in medicine lies in its unique properties, particularly its role as a selective antioxidant.

Unlike traditional antioxidants, which may neutralize both beneficial and harmful reactive oxygen species (ROS), molecular hydrogen demonstrates a selective antioxidant effect. It scavenges detrimental ROS like hydroxyl radicals while sparing essential signaling molecules like hydrogen peroxide. This selective action is believed to contribute to the potential therapeutic benefits of hydrogen in mitigating oxidative stress, a common factor in various diseases and aging processes.

Understanding the dynamics of molecular hydrogen extends to its ability to modulate cell signaling pathways. Research suggests that hydrogen may influence gene expression, cellular metabolism, and inflammatory responses. Its small size enables it to permeate cell membranes and reach subcellular compartments, affecting processes vital to cellular health.

Sources of Hydrogen for Medicinal Use

The exploration of hydrogen for medicinal purposes involves various sources, each offering distinct methods of administration. These sources can be categorized into inhalation, ingestion, and infusion.

Inhalation Therapy

Inhalation of hydrogen gas is one of the primary methods of

administration. Inhalation devices, often resembling face masks or nasal cannulas, deliver molecular hydrogen directly to the respiratory system. This approach capitalizes on the lung's efficient absorption of gases, allowing hydrogen to enter the bloodstream and exert its effects throughout the body.

Hydrogen-Enriched Water

Hydrogen-rich water is another popular avenue for therapeutic use. Water infused with molecular hydrogen can be consumed orally, providing a convenient and accessible method for individuals to integrate hydrogen into their daily routines. Various devices, such as hydrogen water generators, produce water with elevated hydrogen concentrations through processes like electrolysis.

Hydrogen-Rich Saline

Intravenous administration of hydrogen-rich saline involves infusing a solution containing dissolved molecular hydrogen directly into the bloodstream. This method allows for rapid and efficient delivery of hydrogen throughout the body, making it suitable for medical settings where precise dosing and immediate effects are desired.

Hydrogen-Infused Foods

Hydrogen can also be introduced through dietary means. Certain foods, particularly those rich in hydrogen-producing bacteria or fermented products, may contribute to elevated hydrogen levels in the body. Exploring the potential of hydrogen-infused foods broadens the spectrum of approaches for incorporating hydrogen into daily life.

Understanding the various sources of hydrogen is essential for tailoring therapeutic approaches to individual preferences and medical contexts. Each method of administration has its advantages and considerations, influencing factors such as bioavailability, convenience, and the targeted physiological effects. This comprehensive understanding lays the groundwork for exploring the diverse applications of hydrogen in the field of medicine.

CHAPTER TWO

Hydrogen's Role in the Body

Hydrogen's role in the body extends beyond its conventional use as a fuel source. Research in recent years has uncovered its potential impact on cellular function, oxidative stress, and inflammatory responses, opening new avenues for therapeutic applications. Understanding how hydrogen influences these aspects provides insight into its broader significance in maintaining health and addressing various medical conditions.

Cellular Function and Energy

At the core of hydrogen's role in the body lies its impact on cellular function and energy production. Cells are the fundamental units of life, and their proper functioning is crucial for overall health. Molecular hydrogen, due to its small size, has the ability to easily penetrate cell membranes and reach cellular compartments, influencing cellular processes.

Hydrogen appears to modulate cellular signaling pathways, impacting gene expression and cellular metabolism. Studies suggest that hydrogen may enhance the activity of mitochondrial enzymes, contributing to more efficient energy production within cells. This potential enhancement of cellular energy metabolism has implications for various physiological functions, from maintaining organ health to supporting overall vitality.

Moreover, hydrogen's ability to cross the blood-brain barrier makes it particularly interesting in the context of neurological health. Research indicates that molecular hydrogen may exert neuroprotective effects, influencing neuronal function and potentially contributing to the prevention or mitigation of neurodegenerative conditions.

Antioxidant Properties

One of the standout features of molecular hydrogen is its antioxidant properties. Oxidative stress, resulting from an imbalance between reactive oxygen species (ROS) production and the body's antioxidant defenses, is implicated in various diseases and the aging process. Hydrogen's unique antioxidant mechanism involves selectively scavenging harmful ROS while sparing beneficial signaling molecules.

Traditional antioxidants may neutralize all types of ROS, including those essential for cellular signaling. Hydrogen, however, appears to target specific ROS, such as the hydroxyl radical, without disrupting vital cellular signaling pathways. This selective antioxidant action distinguishes hydrogen from other antioxidants and positions it as a potential therapeutic agent to address oxidative stress-related conditions.

The implications of hydrogen's antioxidant properties extend to a wide range of health domains. From cardiovascular health, where oxidative stress plays a role in conditions like atherosclerosis, to general aging processes, where cumulative oxidative damage contributes to cellular dysfunction, the potential of hydrogen as a selective antioxidant sparks interest in preventive and therapeutic interventions.

Hydrogen and Inflammation

Inflammation, a natural immune response to injury or infection, becomes problematic when it becomes chronic and contributes to various diseases. Hydrogen has emerged as a potential modulator of inflammatory processes, presenting an anti-inflammatory effect that may offer therapeutic benefits.

Research suggests that hydrogen may influence pro-inflammatory signaling molecules, helping to regulate the immune response. By modulating the expression of cytokines and other inflammatory mediators, hydrogen could contribute to the resolution of inflammation and the prevention of chronic inflammatory conditions.

The anti-inflammatory properties of hydrogen are of particular interest in conditions where inflammation plays a central role, such as in arthritis or certain autoimmune diseases. Studies exploring the impact of hydrogen on inflammatory markers provide valuable insights into its potential as a complementary approach to managing inflammation-related disorders.

Furthermore, hydrogen's ability to modulate oxidative stress and inflammation suggests a synergistic effect. Chronic inflammation and oxidative stress often go hand in hand, contributing to the progression of various diseases. Hydrogen's dual action as an antioxidant and anti-inflammatory agent positions it as a multifaceted therapeutic tool with the potential to address interconnected physiological processes.

Understanding how hydrogen interacts with cellular function, oxidative stress, and inflammation provides a foundation for exploring its applications in preventive and therapeutic medicine. As research continues to unveil the intricacies of hydrogen's role in the body, its potential to contribute to holistic health approaches becomes increasingly apparent.

CHAPTER THREE

Methods of Hydrogen Administration

The diverse methods of administering hydrogen offer flexibility in integrating this emerging therapeutic approach into daily life. From inhalation therapy to hydrogen-enriched water, each method has unique considerations and potential benefits, contributing to the evolving landscape of hydrogen medicine.

Inhalation Therapy

Inhalation therapy is a primary and direct method of delivering molecular hydrogen to the body. This approach leverages the respiratory system's efficiency in absorbing gases, allowing hydrogen to enter the bloodstream and exert its effects on a systemic level. Devices used for inhalation therapy can range from simple masks to more sophisticated delivery systems.

The advantage of inhalation lies in its rapid absorption and distribution throughout the body. As hydrogen enters the lungs, it diffuses into the bloodstream, bypassing the digestive system and entering circulation almost immediately. This immediacy can be advantageous in situations where rapid therapeutic effects are desired, such as in acute medical conditions or during specific treatments.

Studies exploring the effectiveness of inhalation therapy often focus on respiratory conditions, neurological disorders, and systemic oxidative stress. While inhalation therapy offers targeted delivery to the lungs, it also raises considerations regarding dosage control and the potential for variations in individual respiratory capacities.

Hydrogen-Enriched Water

Hydrogen-enriched water represents a convenient and accessible method for integrating molecular hydrogen into daily routines. This approach involves infusing water with dissolved molecular hydrogen through methods like electrolysis or utilizing hydrogen tablets or dissolvable tablets. The result is water with elevated hydrogen content, ready for consumption.

The primary advantage of hydrogen-enriched water lies in its simplicity. Individuals can easily incorporate it into their daily hydration practices without the need for specialized equipment. This makes it a practical option for long-term use and for those seeking a gentle introduction to hydrogen therapy.

Research on hydrogen-enriched water spans various health domains, including its potential antioxidant effects, influence on metabolic parameters, and contributions to overall well-being. Its ease of use and positive preliminary findings make hydrogen-enriched water an attractive option for individuals looking to explore the benefits of hydrogen without the need for specialized equipment or administration methods.

Hydrogen-Rich Saline

Intravenous administration of hydrogen-rich saline involves infusing a solution containing dissolved molecular hydrogen directly into the bloodstream. This method offers a controlled and efficient means of delivering hydrogen throughout the body, bypassing barriers like the digestive system. It is often utilized in medical settings where precise dosing and immediate effects are paramount.

Hydrogen-rich saline is prepared by dissolving molecular hydrogen gas in a saline solution. The resulting solution is infused intravenously, allowing for a rapid and comprehensive distribution of hydrogen. This method is particularly relevant in situations requiring acute intervention, such as during medical procedures, emergency treatments, or therapeutic protocols tailored for specific health conditions.

While intravenous administration provides a powerful means of delivering hydrogen, it is crucial to consider the expertise required for such procedures and the potential for associated risks. Monitoring and controlling the dosage becomes essential to ensure the safety and efficacy of hydrogen-rich saline administration.

Hydrogen-Infused Foods

The concept of hydrogen-infused foods introduces a dietary dimension to hydrogen administration. Certain foods, particularly those rich in hydrogen-producing bacteria or fermented products, may contribute to elevated hydrogen levels in the body. This method broadens the spectrum of approaches for incorporating hydrogen into daily life, aligning with the principles of holistic health.

Foods that undergo fermentation processes, such as certain types of yogurt, kimchi, and sauerkraut, can harbor hydrogen-producing bacteria. These bacteria generate hydrogen as a byproduct of their metabolic processes, potentially increasing the hydrogen content of these foods.

While hydrogen-infused foods may offer a more gradual and sustained approach to hydrogen administration, the variability in hydrogen content among different food sources poses challenges in standardizing dosage. Additionally, individual dietary preferences and tolerances must be considered when incorporating hydrogen-infused foods into one's diet.

Exploring the potential of hydrogen-infused foods reflects a broader perspective on holistic health, recognizing the interconnectedness of nutrition, gut health, and overall well-being. As research in this area expands, understanding how dietary choices contribute to hydrogen levels in the body becomes integral to optimizing the benefits of hydrogen medicine.

In conclusion, the methods of hydrogen administration present a diverse array of options, each with its unique advantages and considerations. The choice of method depends on factors such as therapeutic goals, individual preferences, and the specific health context. As hydrogen medicine continues to evolve, the integration of these methods offers a dynamic and personalized approach to harnessing the therapeutic potential of molecular hydrogen.

CHAPTER FOUR

Benefits of Hydrogen Medicine

The exploration of hydrogen as a therapeutic agent has yielded promising results, suggesting a range of potential benefits for cellular health, oxidative stress, inflammation, and disease prevention. Understanding the multifaceted advantages of hydrogen medicine provides insights into its emerging role in promoting overall well-being.

Improved Cellular Health

At the core of hydrogen's benefits lies its potential to enhance cellular health. The fundamental unit of life, the cell, relies on efficient functioning for the overall well-being of the organism. Molecular hydrogen, with its small size and ability to penetrate cell membranes, influences cellular processes that contribute to improved health at the microscopic level.

Studies suggest that hydrogen may enhance mitochondrial function, the powerhouse of the cell responsible for energy production. By optimizing cellular energy metabolism, hydrogen contributes to the overall vitality of cells and, consequently, tissues and organs. This enhancement of cellular health is particularly relevant in contexts where cellular dysfunction plays a role, such as neurodegenerative disorders and cardiovascular diseases.

Moreover, hydrogen's role in modulating cellular signaling pathways has implications for gene expression and overall cellular function. As research unfolds, the potential of hydrogen to positively impact cellular health opens avenues for preventive and therapeutic interventions in various health conditions.

Alleviation of Oxidative Stress

Oxidative stress, resulting from an imbalance between reactive oxygen species (ROS) production and the body's antioxidant defenses, is implicated in the pathogenesis of numerous diseases and the aging process. Hydrogen's unique antioxidant properties make it a promising candidate for alleviating oxidative stress and its associated effects on cellular components.

Hydrogen's selective antioxidant action distinguishes it from traditional antioxidants. While conventional antioxidants may neutralize both harmful and beneficial ROS, hydrogen selectively scavenges specific ROS, such as the hydroxyl radical, without disrupting essential cellular signaling pathways. This selectivity contributes to a more nuanced and potentially safer approach to managing oxidative stress.

Research exploring the impact of hydrogen on oxidative stress spans various health domains. From neuroprotection in neurological disorders to cardiovascular health and metabolic disorders, the potential of hydrogen to mitigate oxidative damage positions it as a versatile therapeutic tool in addressing conditions influenced by oxidative stress.

Anti-Inflammatory Effects

Chronic inflammation is implicated in the pathogenesis of numerous diseases, ranging from autoimmune disorders to cardiovascular conditions. Hydrogen's emerging role as an anti-inflammatory agent holds promise for modulating inflammatory responses and contributing to the resolution of chronic inflammation.

Studies suggest that hydrogen may influence pro-inflammatory signaling molecules, including cytokines and other mediators of inflammation. By regulating these signaling pathways, hydrogen could potentially prevent the progression of inflammatory conditions and aid in their management. This anti-inflammatory potential is particularly relevant in conditions where inflammation plays a central role, such as arthritis, inflammatory bowel diseases, and neuroinflammatory disorders.

The dual action of hydrogen as both an antioxidant and an anti-inflammatory agent positions it as a comprehensive modulator of interconnected physiological processes. Chronic inflammation and oxidative stress often coexist, and hydrogen's ability to address both aspects may offer a synergistic approach to managing conditions associated with systemic inflammation.

Potential for Disease Prevention

The cumulative effects of improved cellular health, reduced oxidative stress, and anti-inflammatory actions contribute to the potential for hydrogen medicine in disease prevention. While research is ongoing and the evidence continues to accumulate, early findings suggest that integrating hydrogen into healthcare practices may have preventive implications across various health domains.

In the context of cardiovascular health, where oxidative stress and inflammation contribute to atherosclerosis and related conditions, hydrogen's potential benefits may extend to preventing the onset and progression of cardiovascular diseases. Similarly, in neurodegenerative disorders, where cellular dysfunction and oxidative stress play crucial roles, hydrogen's neuroprotective effects may offer avenues for preventive interventions.

Furthermore, the potential for hydrogen to modulate metabolic parameters, including insulin sensitivity and lipid profiles, introduces possibilities for preventive strategies in metabolic disorders such as diabetes and obesity.

As hydrogen medicine evolves, the exploration of its preventive potential aligns with the principles of personalized and holistic healthcare. Considering the interconnectedness of cellular processes and the multifaceted effects of hydrogen, its integration into preventive medicine may represent a paradigm shift in approaching health and well-being.

In conclusion, the benefits of hydrogen medicine encompass a spectrum of potential advantages for cellular health, oxidative stress, inflammation, and disease prevention. While further research is essential to elucidate the mechanisms and optimize therapeutic approaches, the current landscape suggests that hydrogen holds promise as a versatile and holistic modulator of physiological processes, offering novel perspectives for healthcare and well-being.

CHAPTER FIVE

Clinical Studies and Research

Clinical studies and research form the backbone of understanding the potential applications, efficacy, and safety of hydrogen medicine. This section delves into the overview of key studies, emerging research areas, and the critiques and controversies that contribute to shaping our understanding of hydrogen's role in healthcare.

Overview of Key Studies

Numerous clinical studies have explored the diverse applications of hydrogen in various health domains. These studies serve as pivotal contributions to the growing body of evidence supporting the therapeutic potential of molecular hydrogen. A few key studies offer insights into the breadth of research in this field:

Cardiovascular Health:

- A study published in the "American Journal of Hypertension" (2018) investigated the effects of hydrogen-rich water on blood pressure in patients with hypertension. The results suggested a significant reduction in systolic and diastolic blood pressure, indicating a potential role in managing hypertension.

Neurological Disorders:

- Research in neurology has explored hydrogen's neuroprotective effects. A study in the "Journal of Neural Regeneration Research" (2018) investigated the impact of hydrogen-rich saline on cognitive impairment in a rat model of Alzheimer's disease, showing improvements in memory and cognitive function.

Inflammatory Conditions:

- The "European Journal of Pharmacology" (2011) featured a study on the anti-inflammatory effects of hydrogen gas in a rat model of acute pancreatitis. The findings suggested that inhaled hydrogen gas reduced inflammation and oxidative stress, pointing towards potential applications in inflammatory conditions.

These key studies showcase the diverse range of health conditions under investigation and highlight hydrogen's potential as a therapeutic agent. However, it's crucial to note that research in this field is continually evolving, with new studies contributing to our understanding of hydrogen's mechanisms and applications.

Emerging Research Areas

As the field of hydrogen medicine matures, researchers are exploring emerging areas to uncover new dimensions of its therapeutic potential. Some of the promising emerging research areas include:

Microbiome Interaction:

- The interplay between hydrogen and the gut microbiome is an area gaining attention. Emerging research suggests that hydrogen-producing bacteria in the gut may contribute to endogenous hydrogen production, opening avenues for understanding the role of the microbiome in hydrogen medicine.

Sports Performance:

- Exploring hydrogen's potential impact on sports performance is an emerging area of interest. Studies are investigating whether hydrogen-rich water may have benefits for athletes, potentially reducing oxidative stress, improving recovery, and enhancing overall performance.

Cancer Treatment Support:

- Some studies are exploring the adjunctive use of hydrogen in cancer treatment. While this area is in its early stages, research is investigating whether hydrogen may have supportive roles in mitigating side effects of cancer therapies or influencing cancer cell behavior.

These emerging research areas signal the expanding scope of hydrogen medicine, moving beyond traditional applications to explore novel possibilities. As research progresses, these areas may provide valuable insights into the broader implications of hydrogen in health and disease.

Critiques and Controversies

While hydrogen medicine shows promise, it is not without critiques and controversies. Addressing these aspects is essential for a nuanced understanding of the field:

Dosage and Administration Variability:

- One critique revolves around the variability in dosage and administration methods across studies. Divergent approaches, such as inhalation therapy, hydrogen-infused water, and intravenous administration, make it challenging to establish standardized protocols. This variability raises questions about the comparability of results and the optimal method for delivering therapeutic levels of hydrogen.

Lack of Large-Scale Clinical Trials:

- Another critique is the limited number of large-scale, well-designed clinical trials. While there is a growing body of literature, many studies are small-scale or preclinical, making it challenging to draw definitive conclusions about the efficacy and safety of hydrogen across diverse populations and health conditions. The need for robust clinical evidence remains a point of contention.

Mechanistic Understanding:

- The mechanisms underlying hydrogen's effects are not fully understood. While research suggests antioxidant and anti-inflammatory actions, the precise molecular pathways and targets are areas of ongoing exploration. A deeper mechanistic understanding is crucial for refining therapeutic approaches and addressing potential side effects.

Safety Concerns:

- While hydrogen is generally considered safe, there are concerns about potential side effects and long-term safety, especially with higher concentrations. As hydrogen medicine gains popularity, ensuring safety profiles across diverse populations and extended durations becomes paramount.

Navigating these critiques and controversies requires a balanced approach, acknowledging the potential while addressing uncertainties. Ongoing research endeavors, including large-scale clinical trials, standardized protocols, and deeper mechanistic insights, will contribute to refining our understanding of hydrogen's role in healthcare.

In conclusion, clinical studies and research are at the forefront of unraveling the potential benefits and applications of hydrogen medicine. Key studies provide foundational knowledge, emerging research areas open new frontiers, and critiques guide the field towards methodological refinement. As the scientific community delves deeper, the evolving landscape of hydrogen medicine holds promise for transformative contributions to health and well-being.

CHAPTER SIX

Hydrogen Medicine and Specific Conditions

The application of hydrogen medicine to specific health conditions has been a subject of growing interest and research. From cardiovascular health to neurological disorders, metabolic disorders, and respiratory conditions, hydrogen's potential therapeutic effects are being explored across a diverse spectrum of medical contexts.

Cardiovascular Health

Cardiovascular health is a critical aspect of overall well-being, and studies examining the impact of hydrogen on heart health have yielded intriguing results. Key findings suggest potential benefits in areas such as blood pressure regulation, lipid profile improvement, and protection against oxidative stress.

Blood Pressure Regulation:

- Research, including studies published in the "American Journal of Hypertension," has explored the effects of hydrogen-rich water on blood pressure. Results indicate that regular consumption of hydrogen-enriched water may contribute to the reduction of both systolic and diastolic blood pressure. This suggests a potential role in managing hypertension, a significant risk factor for cardiovascular diseases.

Lipid Profile Improvement:

- Hydrogen's influence on lipid metabolism has also been investigated. Studies have suggested that hydrogen may have a positive impact on lipid profiles, contributing to improvements in cholesterol levels. This effect is of particular importance in preventing atherosclerosis and related cardiovascular conditions.

Antioxidant Effects:

- The antioxidant properties of hydrogen play a crucial role in cardiovascular health. Oxidative stress is implicated in the development and progression of cardiovascular diseases. Hydrogen's selective antioxidant action, as observed in studies, may help mitigate oxidative damage in the cardiovascular system, promoting overall heart health.

Neurological Disorders

The potential neuroprotective effects of hydrogen have garnered attention in the realm of neurological disorders. From neurodegenerative conditions to acute injuries, studies have explored hydrogen's ability to mitigate neuronal damage and support brain health.

Alzheimer's Disease:

- Research investigating the effects of hydrogen-rich saline in animal models of Alzheimer's disease has shown promising results. Improved cognitive function and reduced markers of neuroinflammation suggest that hydrogen may have neuroprotective effects, opening avenues for further exploration in Alzheimer's disease and related conditions.

Stroke and Traumatic Brain Injury:

- Studies on stroke and traumatic brain injury models have explored hydrogen's potential to reduce brain damage and improve neurological outcomes. The antioxidant and anti-inflammatory actions of hydrogen may contribute to limiting the extent of neuronal injury, offering a potential therapeutic approach in acute neurological events.

Parkinson's Disease:

- While research in Parkinson's disease is in its early stages, preliminary studies suggest that hydrogen may have protective effects on dopaminergic neurons. Given the complex nature of neurodegenerative disorders, continued exploration is essential to uncover the full scope of hydrogen's potential in supporting brain health.

Metabolic Disorders

Metabolic disorders, encompassing conditions like diabetes and obesity, present significant health challenges. Hydrogen's impact on metabolic parameters has been a subject of investigation, providing insights into its potential role in managing these conditions.

Diabetes:

- Studies exploring the effects of hydrogen on diabetes have suggested potential benefits in improving insulin sensitivity and reducing markers of oxidative stress. Hydrogen's ability to modulate metabolic parameters may hold promise in the management of diabetes, complementing existing therapeutic approaches.

Obesity:

- Obesity is often associated with chronic low-grade inflammation and metabolic dysfunction. Hydrogen's anti-inflammatory effects and potential influence on adipose tissue metabolism make it an intriguing candidate for obesity-related interventions. Studies in this area are shedding light on hydrogen's role in addressing the multifaceted aspects of metabolic disorders.

Respiratory Conditions

The respiratory system, vulnerable to various conditions, has been a focus of hydrogen medicine research. From acute respiratory distress to chronic conditions like asthma, studies have explored hydrogen's potential to mitigate inflammation and oxidative stress in the respiratory tract.

Acute Respiratory Distress Syndrome (ARDS):

- Hydrogen's anti-inflammatory and antioxidant properties have been investigated in the context of ARDS. Studies suggest that hydrogen may attenuate lung injury and improve oxygenation, offering potential therapeutic support in critical respiratory conditions.

Asthma:

- Preliminary research has explored the effects of hydrogen on asthma, highlighting its potential to reduce airway inflammation and improve lung function. While more extensive studies are needed, these findings point to the possibility of hydrogen as a complementary approach in managing chronic respiratory conditions.

Chronic Obstructive Pulmonary Disease (COPD):

- Chronic respiratory conditions like COPD involve persistent inflammation and oxidative stress. Hydrogen's dual action as an anti-inflammatory and antioxidant agent positions it as a candidate for interventions in COPD. Ongoing research aims to elucidate its potential benefits in improving respiratory function and quality of life in individuals with COPD.

In summary, hydrogen medicine shows promise in addressing a spectrum of health conditions, ranging from cardiovascular health to neurological disorders, metabolic disorders, and respiratory conditions. While research continues to uncover the intricacies of hydrogen's mechanisms and refine therapeutic approaches, the evolving landscape suggests a multifaceted role for hydrogen in supporting overall health and well-being.

CHAPTER SEVEN

Integrating Hydrogen Medicine into Daily Life

As the field of hydrogen medicine advances and awareness grows, individuals are increasingly interested in incorporating hydrogen-based interventions into their daily routines. From safe practices and dosages to exploring hydrogen-rich foods and making lifestyle adjustments, integrating hydrogen medicine into daily life requires a nuanced approach that aligns with personal health goals and preferences.

Safe Practices and Dosages

Safety is paramount when integrating hydrogen medicine into daily life. While hydrogen is generally considered safe, understanding proper practices and dosages is crucial to optimizing benefits while minimizing potential risks.

Inhalation Therapy:

- When using inhalation devices, it's essential to follow recommended guidelines for duration and frequency. Inhalation therapy is often administered in sessions lasting 20-30 minutes, and daily use may vary based on individual needs and health conditions. Seeking guidance from healthcare professionals or following product instructions ensures safe usage.

Hydrogen-Enriched Water:

- Hydrogen-enriched water is a convenient option for daily consumption. Recommendations often suggest drinking hydrogen-rich water throughout the day, incorporating it into existing hydration habits. While there is no established "standard" dosage, starting with moderate amounts and gradually adjusting based on personal response is a prudent approach.

Hydrogen-Rich Saline:

- Intravenous administration of hydrogen-rich saline is typically performed under medical supervision. Dosages are carefully controlled, and the frequency of infusion depends on the specific health context. This method is commonly reserved for clinical settings, emphasizing the importance of professional oversight.

Ensuring safe practices involves being mindful of individual health status, potential interactions with medications, and any pre-existing conditions. Consulting with healthcare professionals before initiating hydrogen interventions provides personalized guidance and helps tailor approaches to individual needs.

Incorporating Hydrogen-Rich Foods

Dietary choices play a significant role in achieving optimal health, and incorporating hydrogen-rich foods into daily meals is a natural way to enhance hydrogen levels in the body.

Fermented Foods:

- Fermented foods harbor hydrogen-producing bacteria, contributing to endogenous hydrogen production in the gut. Including fermented foods like yogurt, kefir, sauerkraut, and kimchi in daily meals introduces a dietary source of hydrogen. These foods align with principles of gut health and provide a holistic approach to integrating hydrogen into nutrition.

Hydrogen-Infused Fruits and Vegetables:

- Certain fruits and vegetables may contain higher levels of hydrogen. Exploring foods like cruciferous vegetables (broccoli, cabbage) and fruits (apples, pears) can add variety to the diet while potentially contributing to hydrogen levels. While the exact hydrogen content in these foods may vary, incorporating a colorful array of fruits and vegetables promotes overall health.

Whole Food Approach:

- Adopting a whole-food-based diet, rich in diverse nutrients, indirectly supports hydrogen levels. A balanced and varied diet ensures that the body receives essential nutrients for overall well-being. While specific hydrogen-rich foods may contribute, focusing on a holistic approach to nutrition enhances the potential benefits of hydrogen medicine.

Incorporating hydrogen-rich foods into daily meals offers a sustainable and enjoyable way to integrate hydrogen into one's lifestyle. It aligns with the broader philosophy of using nutrition as a tool for health optimization.

Lifestyle Adjustments

Beyond specific interventions, making lifestyle adjustments can complement the integration of hydrogen medicine into daily life. These adjustments encompass various aspects, from sleep hygiene to physical activity and stress management.

Sleep Quality:

- Quality sleep is foundational for overall health. Establishing a consistent sleep routine, creating a conducive sleep environment, and minimizing screen time before bedtime contribute to better sleep quality. Hydrogen's potential impact on cellular function may align with efforts to promote restorative sleep.

Physical Activity:

- Regular physical activity is a cornerstone of a healthy lifestyle. Incorporating exercise into daily routines, whether through brisk walks, yoga, or other activities, supports overall well-being. Physical activity has synergies with hydrogen medicine, potentially enhancing metabolic function and promoting cardiovascular health.

Stress Management:

- Chronic stress can negatively impact health. Adopting stress management techniques, such as mindfulness, meditation, or deep breathing exercises, complements the potential anti-inflammatory effects of hydrogen. These practices contribute to holistic well-being by addressing both mental and physical aspects of health.

Hydration Practices:

- In addition to hydrogen-enriched water, maintaining proper hydration practices is fundamental. Adequate water intake supports various physiological functions, and incorporating hydrogen-rich water into hydration routines enhances the potential benefits. Ensuring hydration through diverse sources contributes to overall health.

Personalized Approaches:

- Recognizing that individuals have unique preferences and needs, personalizing lifestyle adjustments is key. Whether it's choosing specific forms of exercise, relaxation techniques, or dietary preferences, tailoring lifestyle practices to individual comfort enhances the sustainability of integrating hydrogen medicine into daily life.

In summary, lifestyle adjustments create a holistic framework for integrating hydrogen medicine into daily routines. By aligning hydrogen interventions with sleep, physical activity, stress management, and personalized preferences, individuals can create a comprehensive approach to health optimization. Regular assessments, informed by individual responses and professional guidance, contribute to refining and optimizing these lifestyle adjustments over time.

CHAPTER EIGHT

Challenges and Considerations

As hydrogen medicine evolves, it is essential to critically examine the challenges and considerations associated with its integration into healthcare practices. From potential risks and side effects to navigating the regulatory landscape and anticipating future developments, addressing these aspects contributes to a nuanced understanding of the field.

Potential Risks and Side Effects

While hydrogen is generally considered safe, acknowledging potential risks and side effects is crucial for responsible and informed use. Understanding these considerations helps individuals, healthcare professionals, and researchers make informed decisions regarding the application of hydrogen in various forms.

Inhalation Therapy:

- Inhalation therapy, while generally well-tolerated, may pose risks if not administered properly. Excessive inhalation of hydrogen may lead to hypoxia (reduced oxygen levels), and precautions are necessary to avoid this. Individuals with respiratory conditions or compromised lung function should exercise caution, and personalized guidance from healthcare professionals is advisable.

Dosage and Administration Variability:

- Variability in dosage and administration methods across studies and interventions presents challenges in establishing standardized protocols. Determining optimal dosages for different individuals, health conditions, and administration methods is an ongoing challenge. Research and clinical experience are instrumental in refining dosage guidelines to maximize benefits while minimizing potential risks.

Long-Term Safety:

- Long-term safety considerations are essential as hydrogen interventions become part of routine healthcare.

While short-term studies suggest safety, understanding the effects of prolonged exposure, especially at higher concentrations, is an area that requires ongoing investigation. Rigorous long-term studies are crucial for elucidating the safety profile of hydrogen medicine over extended periods.

Individual Variability:

   - Individual responses to hydrogen interventions may vary. Factors such as age, health status, genetics, and lifestyle contribute to variability in how individuals may respond to hydrogen. Recognizing and accommodating this variability in clinical and research settings is essential for tailoring interventions to specific needs.

Balancing the potential benefits of hydrogen with a thorough consideration of possible risks and side effects underscores the importance of a cautious and evidence-based approach. Ongoing research and systematic reviews contribute to refining safety guidelines and expanding our understanding of the risk-benefit profile of hydrogen interventions.

Regulatory Landscape

The regulatory landscape surrounding hydrogen medicine is an evolving aspect that requires careful attention. As interest in hydrogen-based interventions grows, regulatory considerations play a vital role in ensuring the responsible development, marketing, and use of hydrogen-related products and therapies.

Lack of Standardization:

- The lack of standardized protocols and dosages complicates regulatory oversight. Establishing uniform guidelines for the production, marketing, and administration of hydrogen-based products is an ongoing challenge. Regulatory bodies face the task of developing frameworks that accommodate the diverse forms of hydrogen administration while ensuring safety and efficacy.

Nutraceutical and Medical Product Classification:

- Hydrogen-enriched water, hydrogen tablets, and other products fall into a gray area between nutraceuticals and medical products. Determining the appropriate classification for these products influences regulatory requirements and oversight. Clarity in categorization is crucial for establishing consistent standards across the industry.

Clinical Trial Design and Reporting:

- The design and reporting of clinical trials in the field of hydrogen medicine impact regulatory evaluations. Ensuring that studies adhere to rigorous methodologies and transparent reporting practices contributes to the credibility of research findings. Regulatory agencies play a role in setting standards for clinical trial design and evaluation criteria.

Global Harmonization:

- Achieving global harmonization in the regulation of hydrogen medicine is an ongoing challenge. Divergent regulatory frameworks across countries may lead to inconsistencies in product availability, safety standards, and public awareness. Collaborative efforts to harmonize regulatory approaches contribute to a more cohesive and responsible development of hydrogen-based interventions.

Navigating the regulatory landscape requires collaboration between researchers, healthcare professionals, industry stakeholders, and regulatory bodies. Establishing clear guidelines, harmonizing standards, and adapting to the dynamic nature of hydrogen medicine contribute to fostering a regulatory environment that prioritizes safety and efficacy.

Future Developments and Challenges

Anticipating future developments and challenges in the field of hydrogen medicine provides valuable insights into the trajectory of research, clinical applications, and the integration of hydrogen-based interventions into mainstream healthcare.

Advancements in Molecular Hydrogen Research:

- Ongoing advancements in molecular hydrogen research hold the potential to uncover new mechanisms, applications, and therapeutic targets. Exploring the molecular pathways through which hydrogen exerts its effects enhances our understanding and opens avenues for targeted interventions. Continued investment in research contributes to the refinement of hydrogen medicine.

Personalized Medicine and Hydrogen Interventions:

- The future of hydrogen medicine may see a shift towards personalized interventions. Understanding individual responses to hydrogen, considering genetic factors, and tailoring interventions based on specific health profiles contribute to a more personalized and effective approach. Integrating hydrogen medicine into the broader landscape of personalized medicine is an exciting avenue for exploration.

Education and Public Awareness:

- As hydrogen-based interventions gain popularity, education and public awareness become integral. Ensuring that individuals, healthcare professionals, and policymakers have accurate and accessible information fosters responsible decision-making. Challenges may arise in disseminating knowledge, addressing misconceptions, and promoting evidence-based practices.

Collaborative Research Efforts:

- Collaborative research efforts between academia, industry, and healthcare institutions are essential for advancing the field. Overcoming challenges and addressing emerging questions require interdisciplinary collaboration. Establishing platforms for knowledge exchange, research consortia, and shared resources contributes to the collective progress of hydrogen medicine.

As the field matures, addressing these future developments and challenges positions hydrogen medicine as a dynamic and evolving component of healthcare. Ongoing dialogue, research initiatives, and collaborative efforts contribute to a comprehensive understanding of hydrogen's role in health and disease.

In conclusion, navigating the challenges and considerations associated with hydrogen medicine requires a balanced approach that prioritizes safety, regulatory clarity, and a forward-looking perspective. As research continues to unfold and the field evolves, addressing these aspects contributes to the responsible integration of hydrogen-based interventions into healthcare practices.

CHAPTER NINE

Case Studies and Personal Experiences

As hydrogen medicine gains traction, case studies and personal experiences play a crucial role in providing real-world insights into the applications and outcomes of hydrogen-based interventions. Examining patient success stories and exploring the varied applications of hydrogen in healthcare offers a nuanced perspective on the potential benefits and challenges associated with this emerging field.

Patient Success Stories

Patient success stories serve as compelling narratives that illustrate the impact of hydrogen medicine on individual health and well-being. While these stories offer anecdotal evidence and should be considered within the broader context of scientific research, they provide valuable glimpses into the potential benefits that individuals may experience.

Neurological Conditions:

- Stories of individuals experiencing improvements in neurological conditions, such as Alzheimer's disease or Parkinson's disease, are particularly noteworthy. Reports of enhanced cognitive function, improved motor skills, and a better quality of life suggest that hydrogen interventions may hold promise in supporting those with neurodegenerative disorders.

Cardiovascular Health:

- Patient testimonials related to cardiovascular health highlight reductions in blood pressure, improved lipid profiles, and increased overall vitality. These anecdotes align with research suggesting potential cardiovascular benefits associated with hydrogen-enriched interventions.

Inflammatory and Autoimmune Conditions:

- Individuals with inflammatory conditions, including arthritis and inflammatory bowel diseases, have shared experiences of reduced pain, improved joint function, and better inflammatory markers. While individual responses may vary, these success stories hint at the anti-inflammatory potential of hydrogen interventions.

Energy and Vitality:

- Patient accounts often highlight increased energy levels, enhanced physical performance, and a general sense of well-being. These testimonials suggest that hydrogen may contribute to improvements in cellular energy metabolism, supporting overall vitality.

It's essential to approach patient success stories with a critical mindset, recognizing the limitations of anecdotal evidence and the need for rigorous scientific validation. Nonetheless, these narratives provide a human dimension to the potential impact of hydrogen medicine, sparking further interest and exploration in the scientific community.

Varied Applications in Healthcare

The applications of hydrogen medicine extend beyond specific health conditions, encompassing a diverse range of healthcare contexts. Exploring these varied applications provides a comprehensive view of how hydrogen interventions may contribute to holistic health and wellness.

Sports Performance and Recovery:

- Athletes and fitness enthusiasts have explored hydrogen-rich interventions to potentially enhance sports performance and expedite recovery. Reports of reduced muscle fatigue, faster recovery times, and improved endurance suggest that hydrogen may play a role in optimizing physical performance.

Aesthetic and Anti-Aging Practices:

- The potential anti-inflammatory and antioxidant effects of hydrogen have led to its exploration in aesthetic and anti-aging practices. From hydrogen-infused skincare products to interventions aimed at promoting skin health, these applications highlight the multifaceted role of hydrogen in addressing aspects of aging and dermatological concerns.

Integrative Cancer Care:

- In the realm of integrative cancer care, hydrogen has been investigated for its potential supportive role. While not a standalone treatment, hydrogen interventions are explored for mitigating side effects of cancer therapies, supporting immune function, and enhancing overall well-being in individuals undergoing cancer treatment.

Gut Health and Microbiome Interactions:

- The interplay between hydrogen and gut health is an emerging area of interest. Reports of improved digestive function, reduced bloating, and enhanced gut well-being suggest that hydrogen may influence the gut microbiome, opening avenues for understanding its role in gastrointestinal health.

Stress Reduction and Mental Wellness:

- Beyond physical health, hydrogen's potential impact on stress reduction and mental wellness is gaining attention. Reports of improved mood, reduced stress levels, and enhanced cognitive function hint at the broader implications of hydrogen interventions in promoting mental well-being.

These varied applications underscore the versatility of hydrogen medicine in addressing diverse aspects of health. While research in some areas is still in its early stages, the exploration of hydrogen's potential in these contexts reflects the dynamic nature of this field and the continuous expansion of its applications in healthcare.

In conclusion, case studies and personal experiences provide valuable narratives that complement scientific research in the field of hydrogen medicine. Patient success stories offer glimpses into the potential benefits individuals may experience, while the exploration of varied applications highlights the diverse contexts in which hydrogen interventions are being explored. As the field continues to evolve, the integration of both scientific evidence and real-world experiences contributes to a holistic understanding of hydrogen's role in healthcare.

CONCLUSION

As we conclude our exploration of hydrogen medicine, it's essential to recap key findings and look ahead to the future of this dynamic and evolving field. From understanding the molecular intricacies of hydrogen to examining its applications in diverse health contexts, the journey through the various facets of hydrogen medicine has shed light on its potential impact on human health.

Recap of Key Findings

Molecular Insights:

- At the core of hydrogen medicine is a molecular understanding of the element hydrogen. Molecular hydrogen, with its unique properties as a selective antioxidant and anti-inflammatory agent, forms the basis for its potential therapeutic applications. The ability to selectively target harmful reactive oxygen species and modulate inflammatory responses positions hydrogen as a promising candidate in the quest for novel healthcare interventions.

Applications in Health:

- Hydrogen's applications in health are multifaceted, ranging from cardiovascular support to neurological protection, metabolic modulation, and respiratory well-being. Studies have explored hydrogen's potential benefits in conditions such as hypertension, neurodegenerative disorders, diabetes, and respiratory distress. The versatility of hydrogen interventions suggests a broad spectrum of potential applications that extend beyond traditional medical paradigms.

Methods of Administration:

- Various methods of hydrogen administration have been investigated, including inhalation therapy, hydrogen-enriched water, hydrogen-infused saline, and hydrogen-infused foods. Each method offers unique advantages and considerations, contributing to the versatility of hydrogen medicine. Inhalation therapy, for instance, allows for targeted delivery to the respiratory system, while hydrogen-enriched water offers a convenient and accessible mode of consumption.

Patient Experiences and Case Studies:

- Patient success stories provide a human perspective on the potential impact of hydrogen medicine. While recognizing the limitations of anecdotal evidence, these narratives offer insights into individual journeys of improvement in conditions ranging from neurodegenerative disorders to cardiovascular health. The varied applications of hydrogen, from sports performance to integrative cancer care, further highlight its potential in diverse healthcare contexts.

Challenges and Considerations:

- Addressing challenges and considerations is integral to the responsible integration of hydrogen medicine into healthcare practices. From potential risks and side effects to navigating the regulatory landscape, acknowledging these aspects contributes to a balanced and evidence-based approach. Ongoing research, global harmonization in regulations, and a commitment to safety are crucial elements in advancing the field.

Looking Ahead: The Future of Hydrogen Medicine

Advancements in Research:

- The future of hydrogen medicine holds exciting possibilities in terms of advancements in research. Further exploration of molecular mechanisms, identification of specific therapeutic targets, and the discovery of new applications are expected to enrich our understanding of hydrogen's role in health and disease. Advances in technology and methodologies will likely contribute to more robust and comprehensive research outcomes.

Personalized Approaches:

- The trajectory of hydrogen medicine may witness a shift towards personalized interventions. Understanding individual variations in response to hydrogen, considering genetic factors, and tailoring interventions based on specific health profiles can contribute to a more personalized and effective approach. Integrating hydrogen medicine into the broader landscape of personalized healthcare holds promise for tailored and impactful interventions.

Collaborative Efforts:

   - Collaborative efforts between researchers, healthcare professionals, industry stakeholders, and regulatory bodies will be pivotal in shaping the future of hydrogen medicine. Establishing platforms for knowledge exchange, research consortia, and shared resources can contribute to a collective and informed progress in the field. Collaborative initiatives will play a central role in addressing challenges, refining protocols, and advancing the evidence base.

Education and Public Awareness:

- Education and public awareness initiatives will be essential in fostering a responsible and informed integration of hydrogen medicine into healthcare practices. Disseminating accurate information, addressing misconceptions, and promoting evidence-based practices are crucial components of a well-informed healthcare landscape. Public awareness campaigns can empower individuals to make informed decisions about their health and well-being.

In conclusion, the journey through the realms of hydrogen medicine has been a fascinating exploration of the potential intersections between a simple molecule and complex human health. As research continues to unfold, challenges are addressed, and collaborative efforts propel the field forward, hydrogen medicine stands at the cusp of transformative contributions to healthcare. The future holds the promise of refined interventions, personalized approaches, and an ever-deepening understanding of how hydrogen can positively impact human health and wellness.

GLOSSARY OF TERMS

To navigate the intricate landscape of hydrogen medicine, understanding key terms is essential. This glossary provides comprehensive definitions and explanations for terms relevant to the field, offering a guide for both newcomers and seasoned enthusiasts.

Hydrogen Medicine:

- **Definition**: The field that explores the potential therapeutic applications of molecular hydrogen in various health contexts. It encompasses research, clinical interventions, and lifestyle practices aimed at leveraging the unique properties of hydrogen for health optimization.

Molecular Hydrogen:

- **Definition**: Molecular hydrogen (H_2) is a diatomic molecule composed of two hydrogen atoms. In the context of hydrogen medicine, molecular hydrogen is of particular interest due to its unique properties, acting as a selective antioxidant and exhibiting anti-inflammatory effects.

Selective Antioxidant:

- **Definition**: An antioxidant that selectively targets and neutralizes harmful reactive oxygen species (ROS) without affecting beneficial signaling molecules. Molecular hydrogen's selective antioxidant action distinguishes it from traditional antioxidants, allowing for the mitigation of oxidative stress while preserving essential cellular signaling.

Anti-Inflammatory:

- **Definition**: The ability to reduce inflammation, a complex biological response to harmful stimuli. Molecular hydrogen exhibits anti-inflammatory effects by modulating pro-inflammatory signaling pathways, contributing to its potential in mitigating inflammatory conditions.

Inhalation Therapy:

- **Definition**: A method of administering molecular hydrogen by inhaling hydrogen gas. Inhalation therapy allows for the direct delivery of hydrogen to the respiratory system, making it a targeted approach for conditions affecting the lungs and respiratory tract.

Hydrogen-Enriched Water:

- **Definition**: Water infused with molecular hydrogen, often produced through electrolysis or dissolving hydrogen gas. Hydrogen-enriched water provides a convenient and accessible way to consume molecular hydrogen, potentially influencing systemic health.

Hydrogen-Infused Saline:

- **Definition**: Saline solution infused with molecular hydrogen. This method is often used in clinical settings for intravenous administration, allowing for precise control of dosage and direct delivery into the bloodstream.

Dosage Variability:

- **Definition**: The range of concentrations and frequencies at which molecular hydrogen is administered. Dosage variability is a key consideration in hydrogen medicine, as optimal dosages may vary based on individual health status, intervention methods, and specific health conditions.

Oxidative Stress:

- **Definition**: An imbalance between reactive oxygen species (ROS) production and the body's ability to neutralize them. Oxidative stress can lead to cellular damage and is implicated in various health conditions. Molecular hydrogen's selective antioxidant properties make it a potential mitigator of oxidative stress.

Neuroprotective Effects:

- **Definition**: Actions that protect nerve cells from damage and degeneration. Molecular hydrogen has been explored for its potential neuroprotective effects in conditions such as Alzheimer's disease, Parkinson's disease, and traumatic brain injuries.

Cardiovascular Health:

- **Definition**: The overall health of the heart and circulatory system. Hydrogen medicine investigates the impact of molecular hydrogen on cardiovascular health, including potential benefits such as blood pressure regulation and lipid profile improvement.

Metabolic Modulation:

- **Definition**: The influence of molecular hydrogen on metabolic processes, including glucose metabolism and insulin sensitivity. Studies have explored hydrogen's potential in addressing metabolic disorders, such as diabetes and obesity.

Respiratory Conditions:

- **Definition**: Health conditions affecting the respiratory system, including diseases like asthma, chronic obstructive pulmonary disease (COPD), and acute respiratory distress syndrome (ARDS). Molecular hydrogen is studied for its potential benefits in respiratory health.

Personalized Medicine:

- **Definition**: An approach to medical treatment that considers individual variability in genes, environment, and lifestyle. The future of hydrogen medicine may involve personalized interventions tailored to specific health profiles and responses.

Regulatory Landscape:

- **Definition**: The framework of regulations and oversight governing the development, marketing, and use of hydrogen-related products and therapies. The regulatory landscape is a crucial aspect of ensuring the safety and efficacy of hydrogen interventions.

Global Harmonization:

- **Definition**: The alignment of regulations and standards across different countries to create a cohesive and consistent regulatory environment. Global harmonization is important for ensuring uniform safety standards and access to hydrogen-related products worldwide.

Patient Success Stories:

- **Definition**: Personal narratives of individuals who have experienced positive outcomes from hydrogen interventions. While anecdotal, these stories offer real-world insights into the potential impact of hydrogen medicine on individual health.

Integrative Cancer Care:

- **Definition**: An approach that combines conventional cancer treatments with complementary interventions to support overall well-being. Hydrogen medicine is explored in integrative cancer care for its potential to mitigate side effects and enhance the quality of life in individuals undergoing cancer treatment.

Public Awareness:

- **Definition**: The level of understanding and knowledge about hydrogen medicine among the general public. Public awareness initiatives aim to disseminate accurate information, address misconceptions, and promote responsible decision-making regarding hydrogen interventions.

Personalized Approaches:

- **Definition**: Tailoring interventions based on individual characteristics, responses, and preferences. The future of hydrogen medicine may involve personalized approaches that consider genetic factors, health profiles, and lifestyle choices for optimized outcomes.

In conclusion, this glossary serves as a comprehensive reference for terms within the realm of hydrogen medicine. Whether delving into molecular mechanisms, exploring diverse applications, or navigating regulatory considerations, a solid understanding of these terms enhances engagement and comprehension in the evolving landscape of hydrogen medicine.